SUPERCAR
COLORING BOOK
FOR KIDS AGE 4-8, 8-12

BY AS PUBLISHING

Copyright © 2020 by As Publishing
All rights reserved. This book or any portion thereof
may not be reproduced or used in any manner whatsoever
without the express written permission of the publisher
except for the use of brief quotations in a book review.

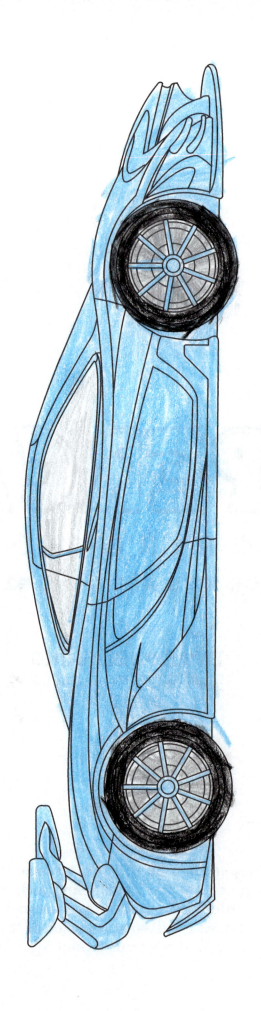

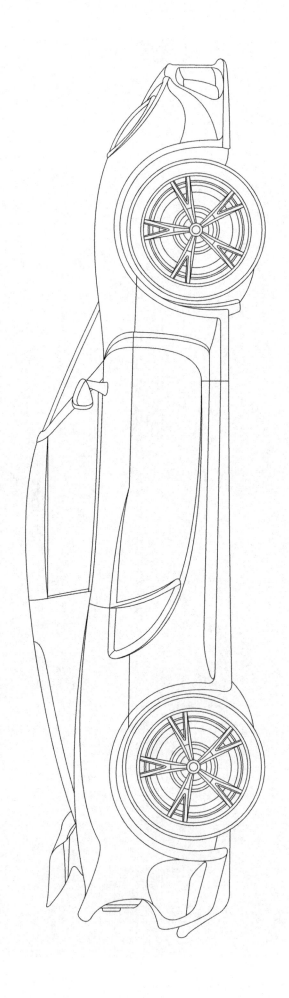

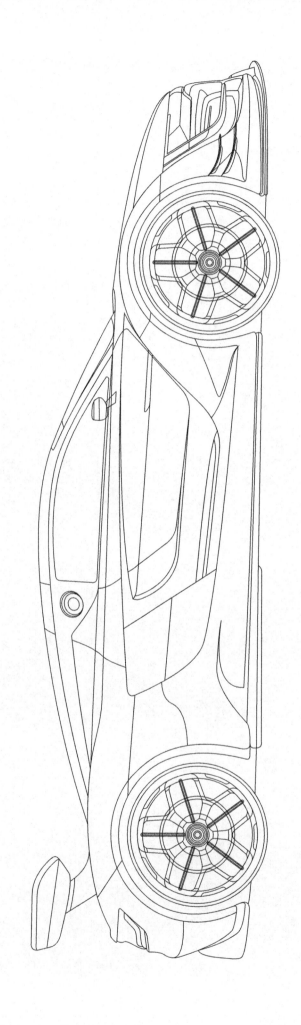

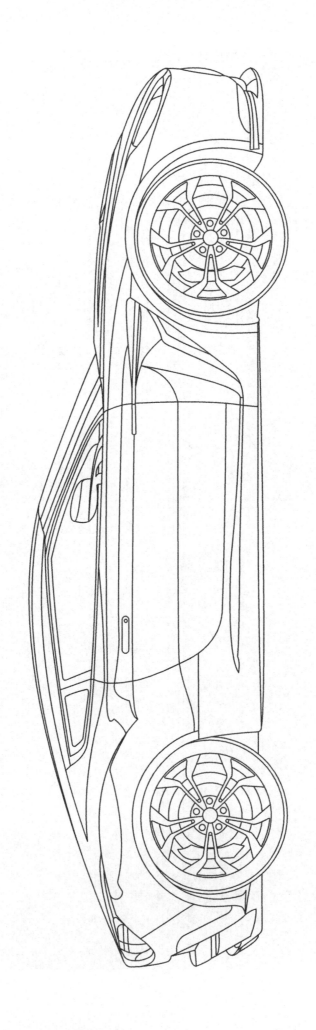

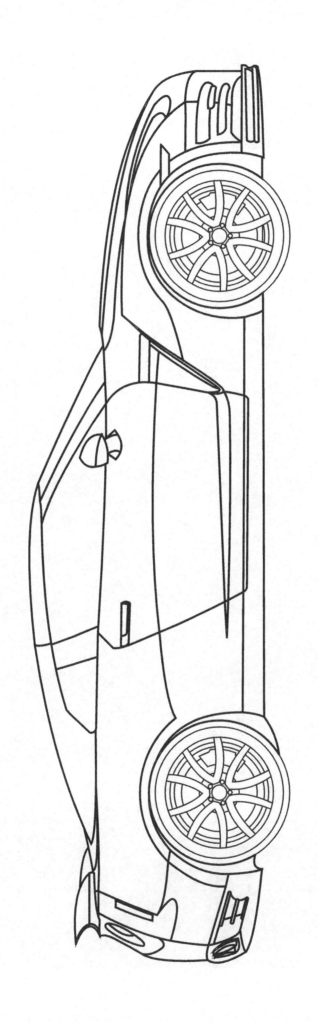

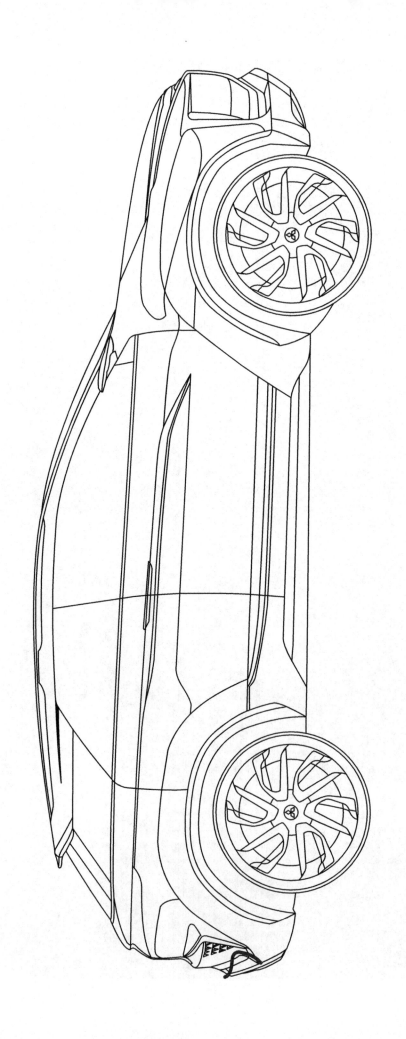

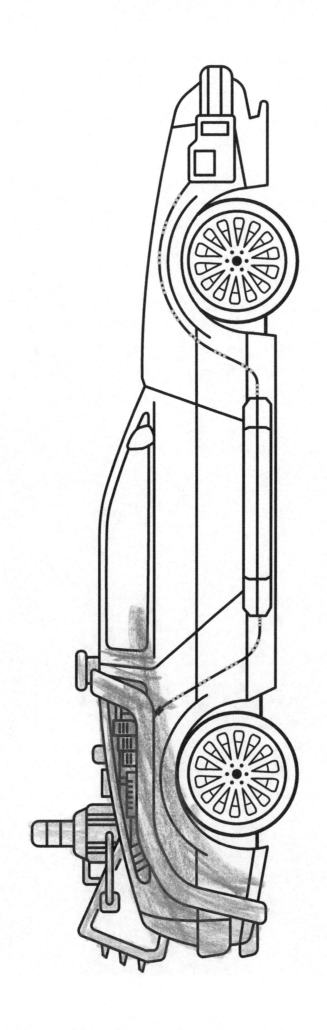

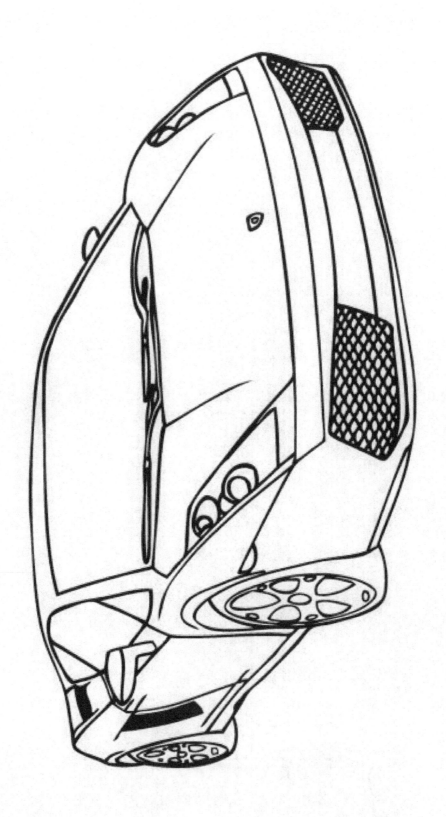

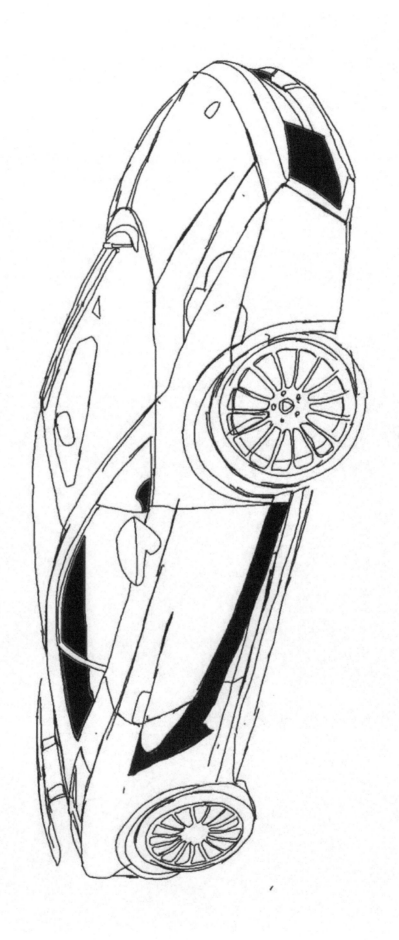

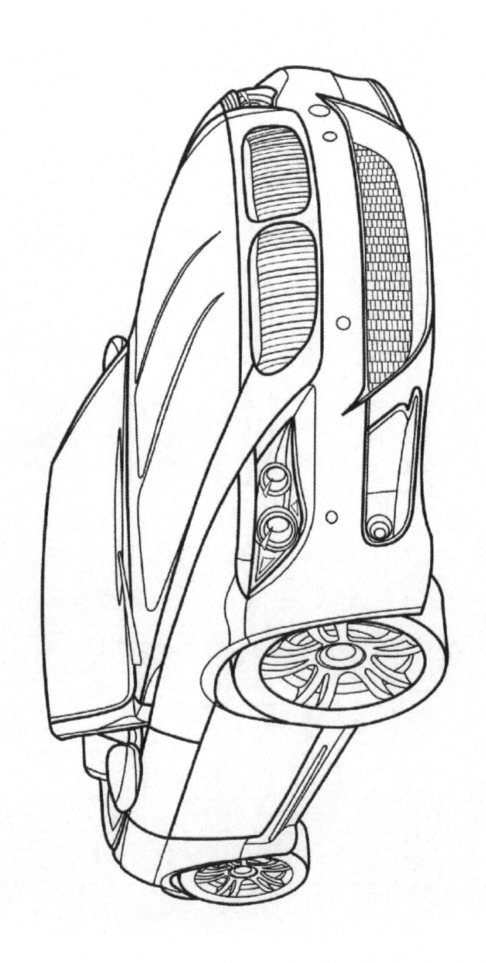

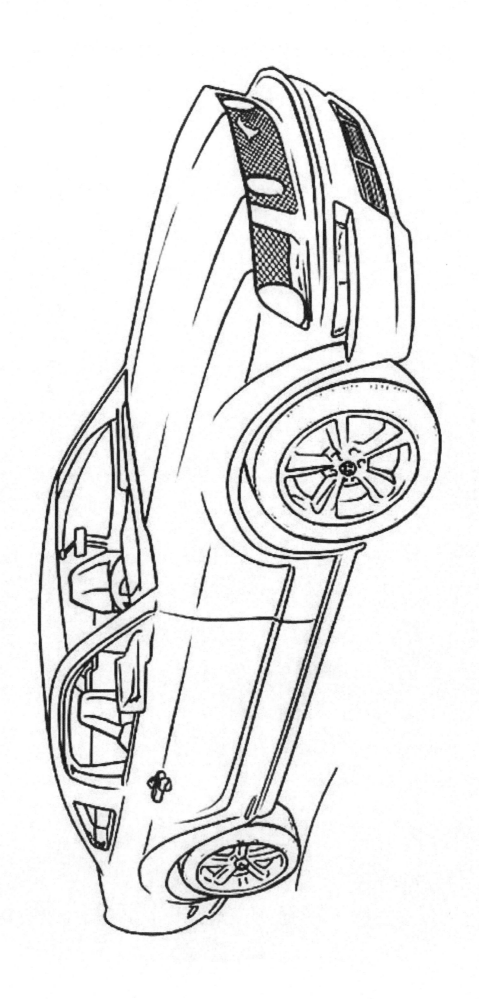

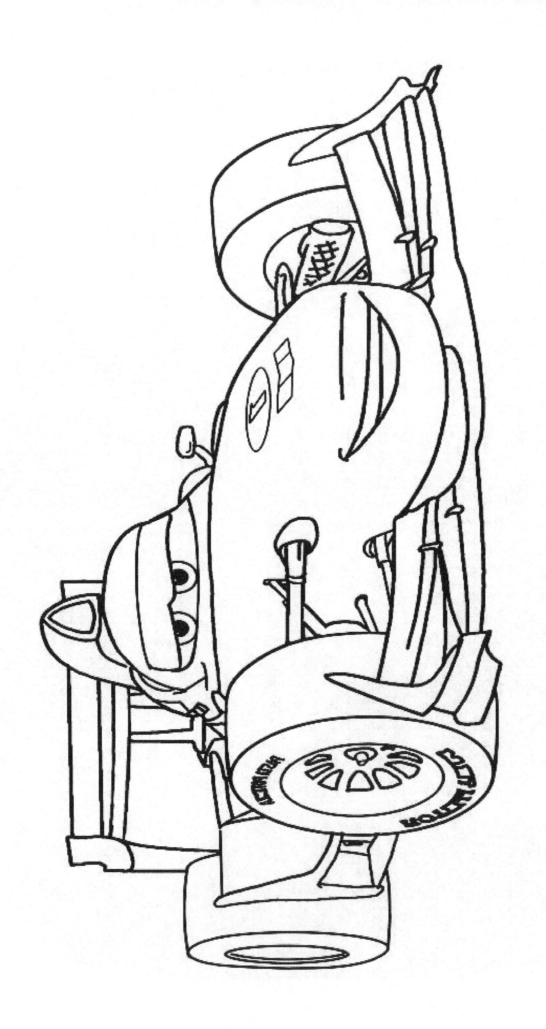

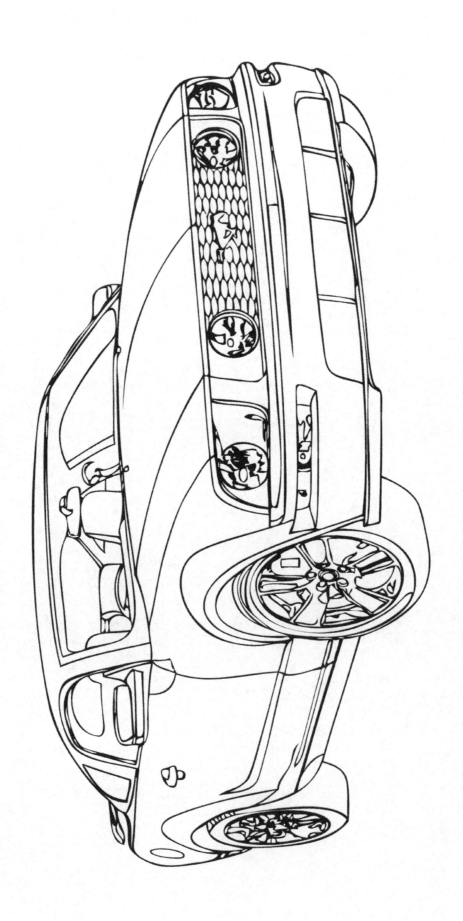

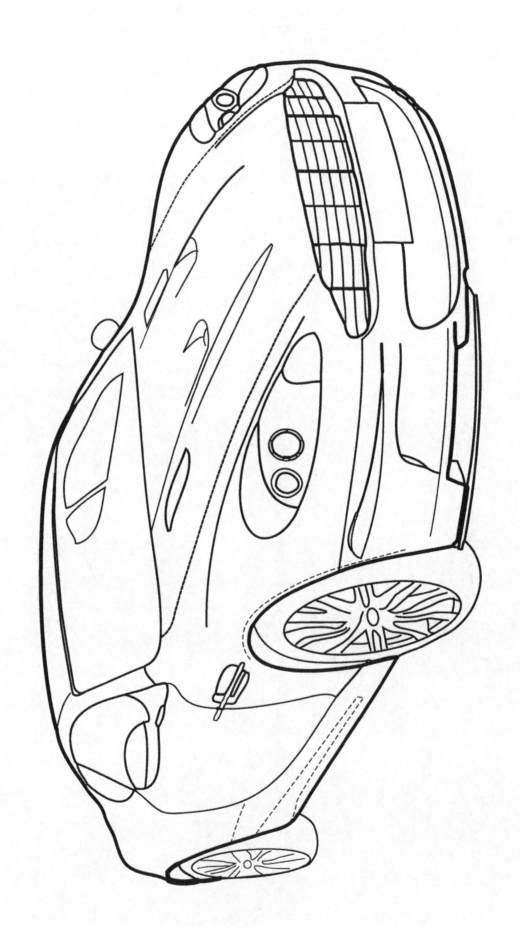

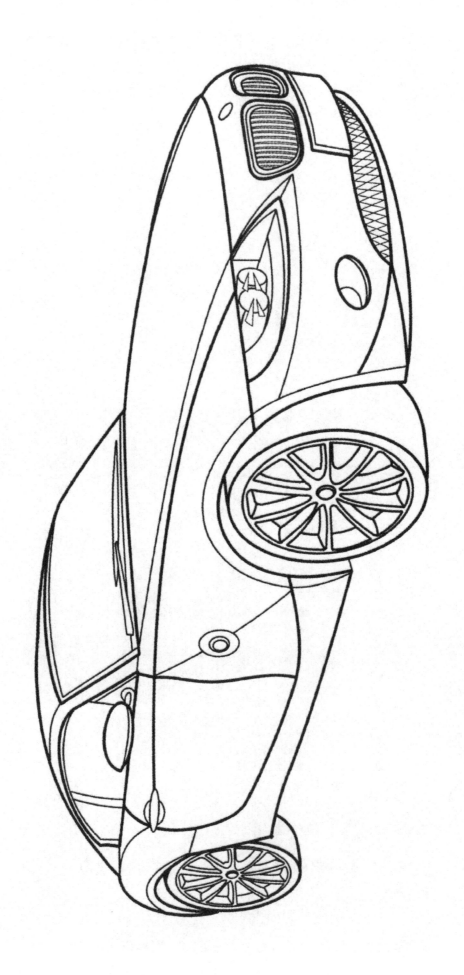

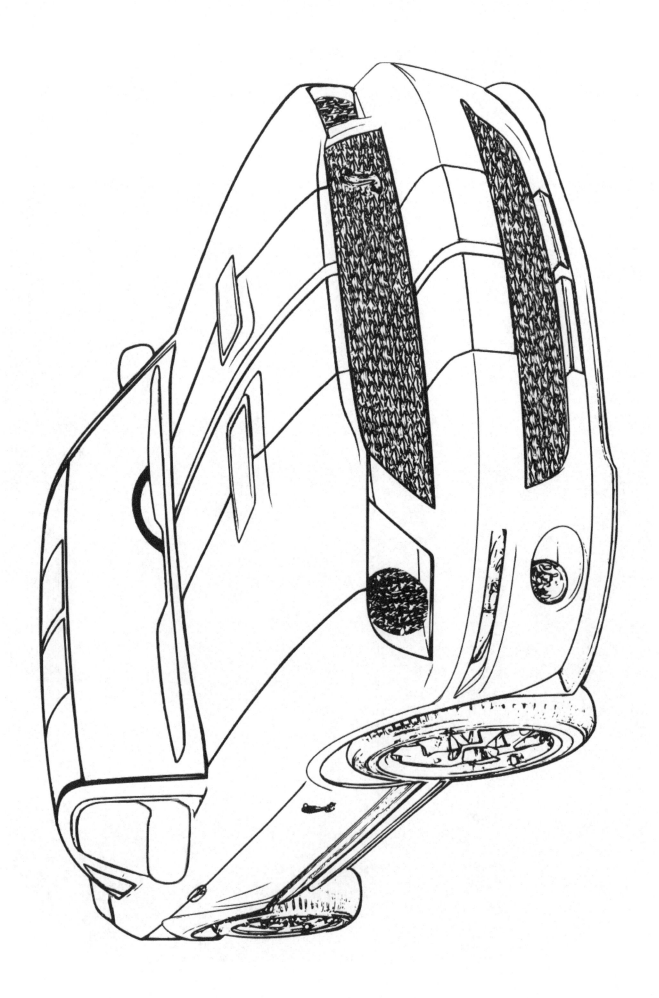

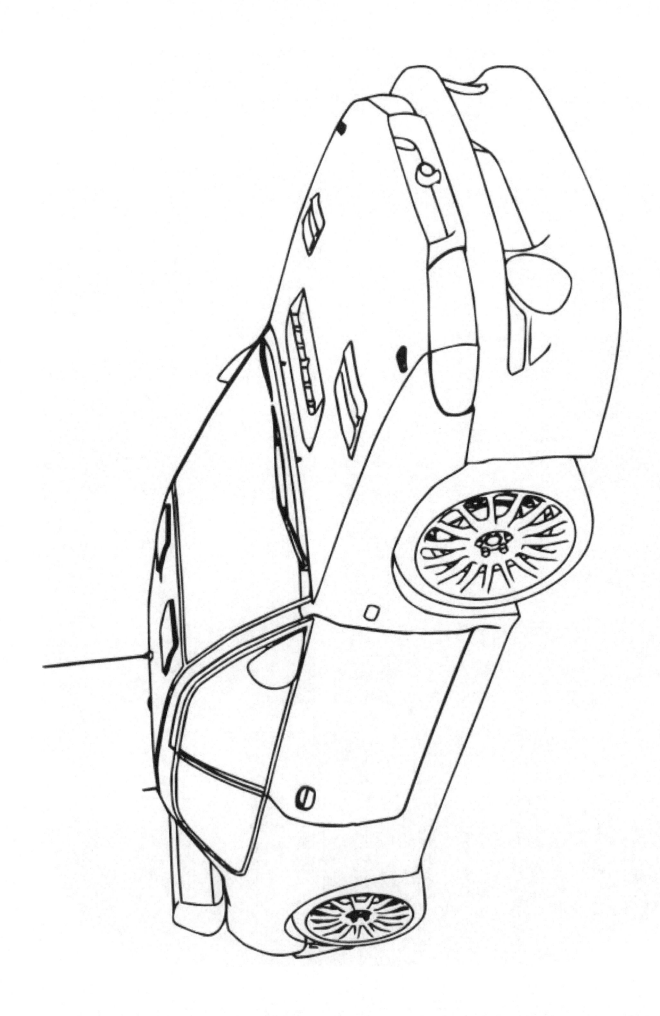

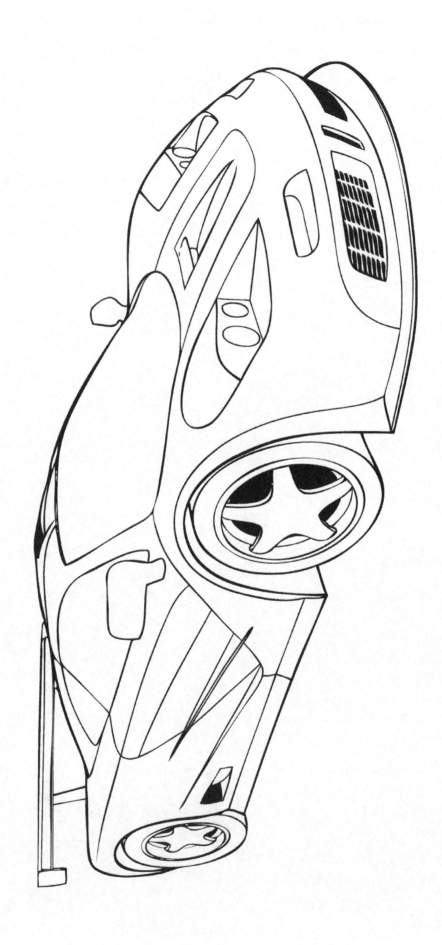

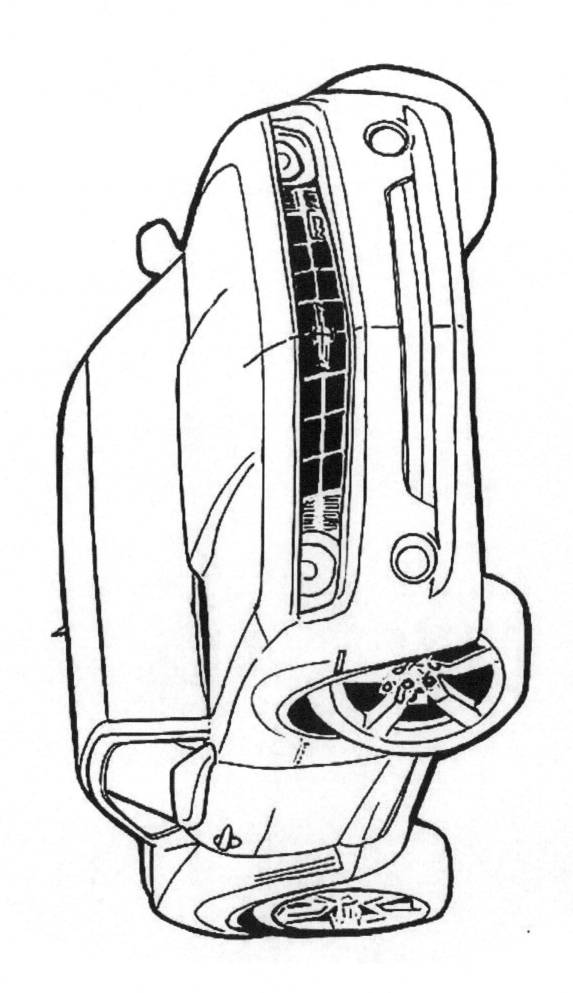

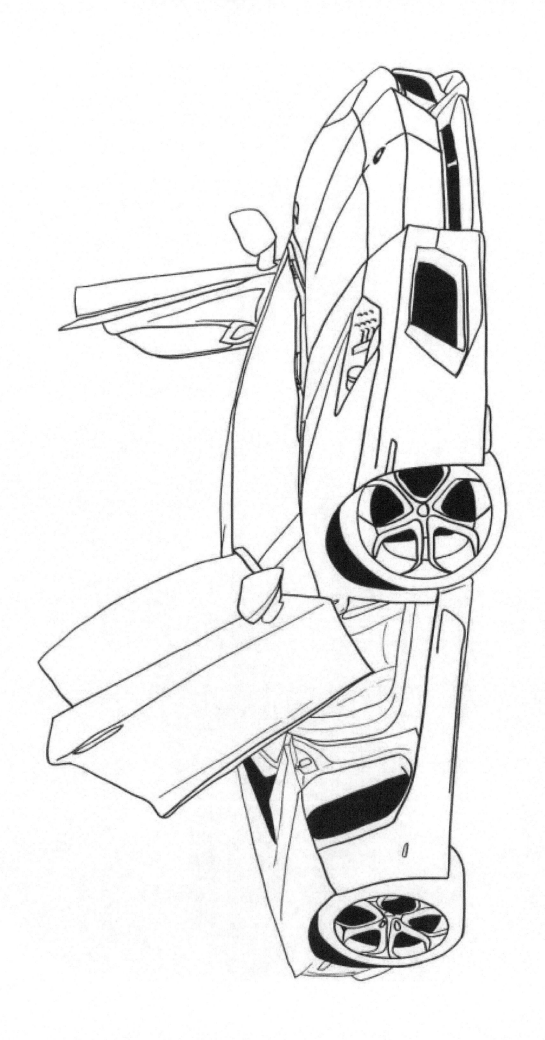

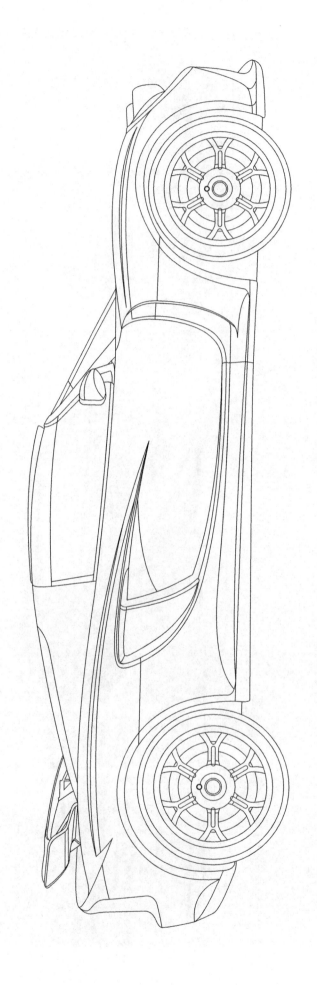

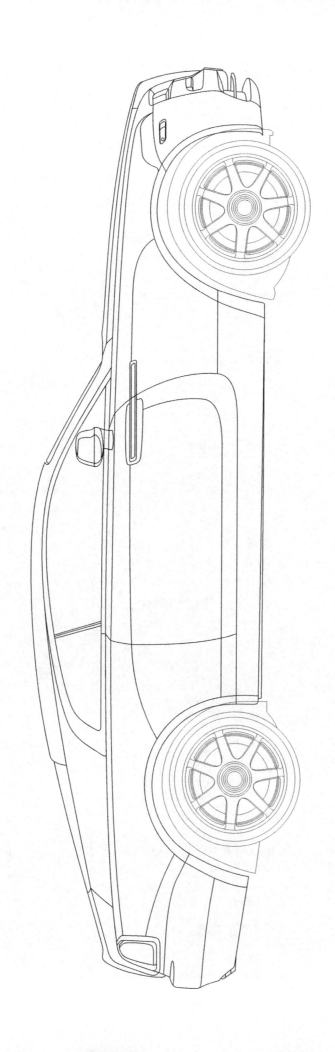

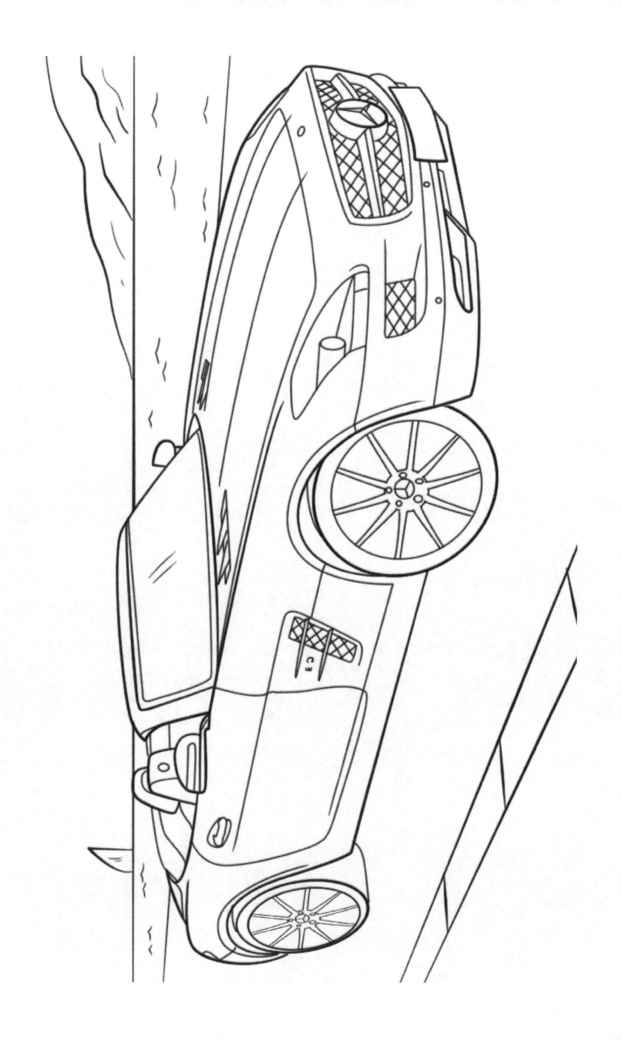

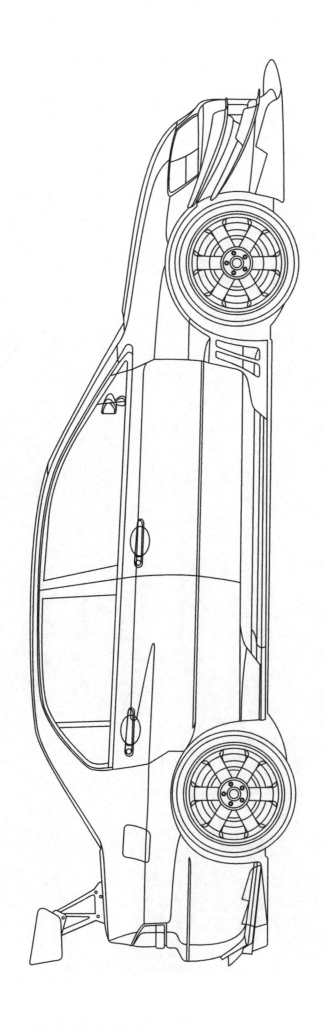

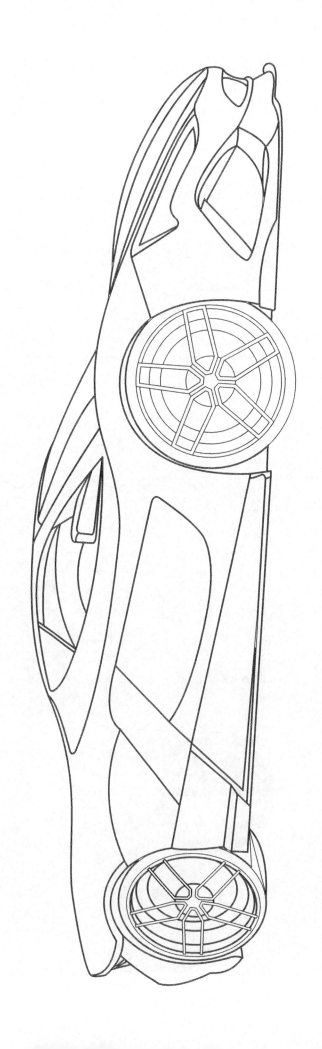

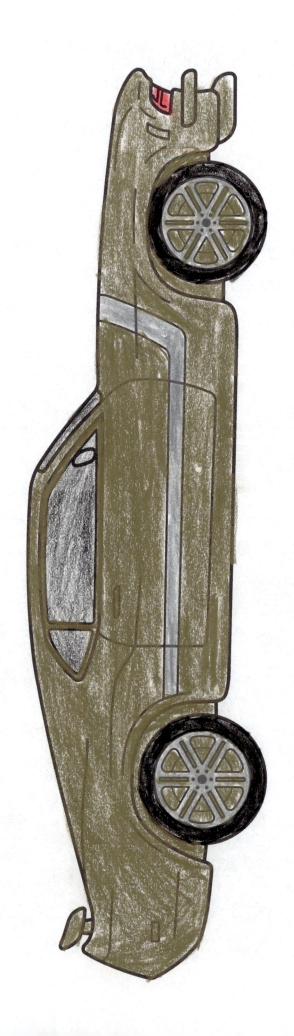

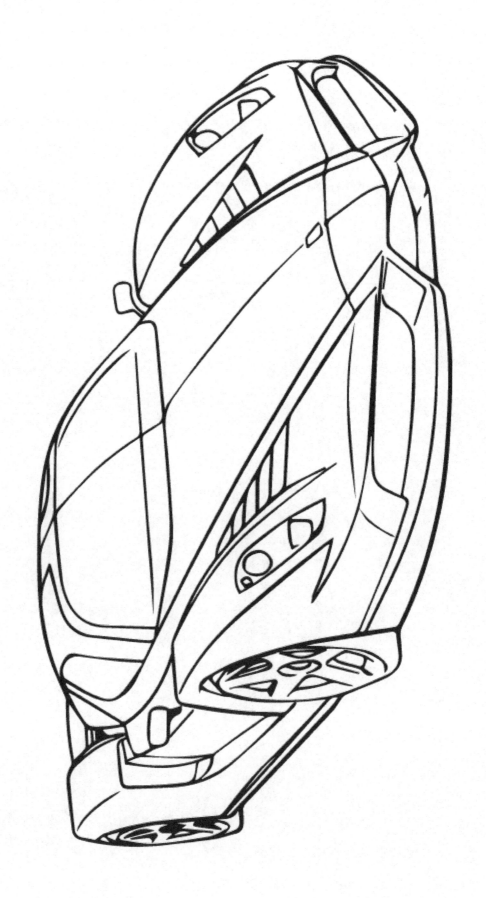

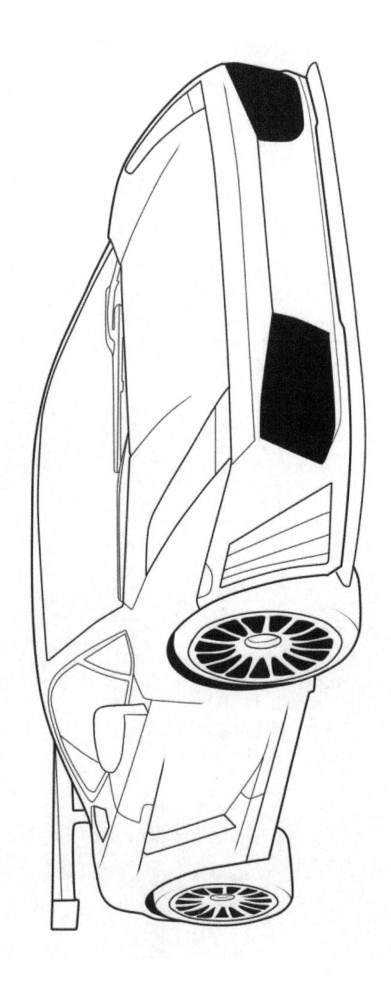

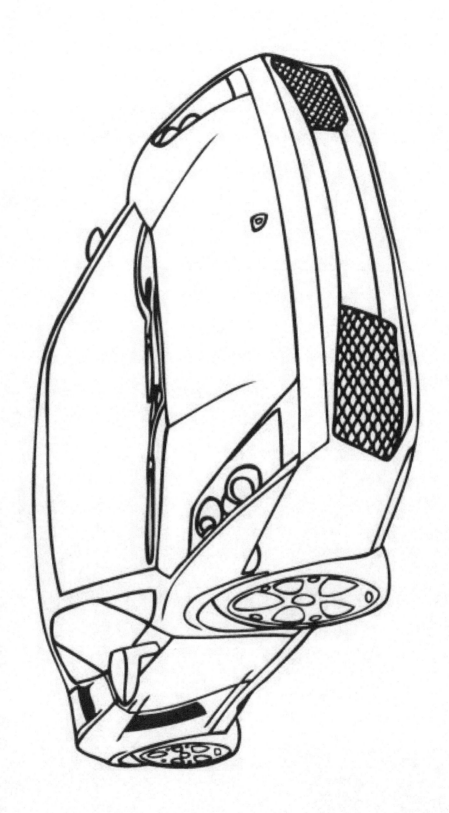

Made in the USA
Monee, IL
30 October 2020